# Fires & Bombs at Three Mills by Brian St...

Three Mills has been a site of milling since ... a distillery had been established. The Hous... Bisson in 1776, between two houses (henc... he lived and the other he had built in 1763 for his son Daniel Junior. Both Daniel Bissons died in 1777 and Three Mills was first leased to Philip Metcalfe and subsequently purchased by him. Throughout 300 years, these buildings have been subject to a number of disasters.

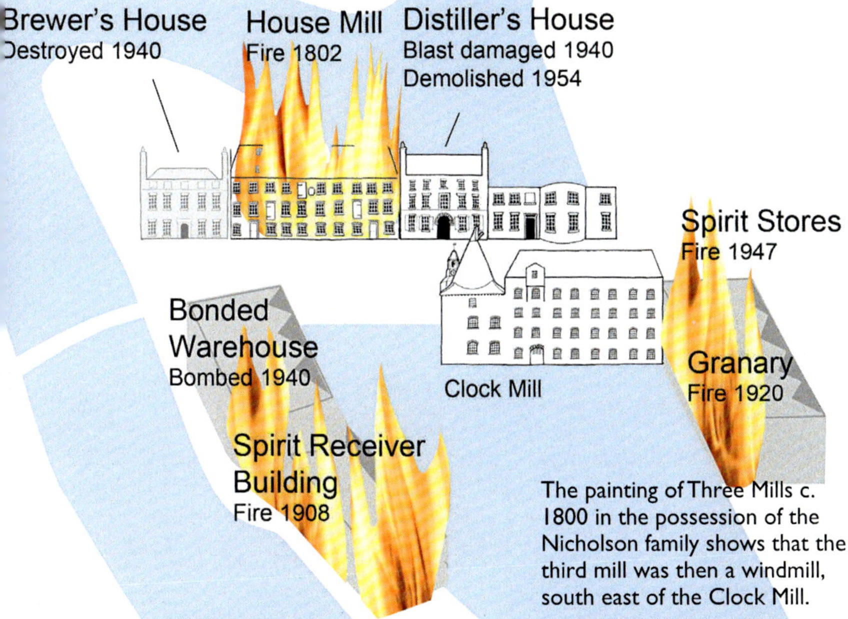

**Brewer's House**
Destroyed 1940

**House Mill**
Fire 1802

**Distiller's House**
Blast damaged 1940
Demolished 1954

**Spirit Stores**
Fire 1947

**Bonded Warehouse**
Bombed 1940

**Clock Mill**

**Granary**
Fire 1920

**Spirit Receiver Building**
Fire 1908

The painting of Three Mills c. 1800 in the possession of the Nicholson family shows that the third mill was then a windmill, south east of the Clock Mill.

In her history of Three Mills, E M Gardner says that the two watermills were valued at £150 each in 1820 and the windmill at £20. She quotes a diarist, Dr Pagenstecher, as recording that the windmill was struck by lightning and wrecked in 1830 and that he gave a picture of it on the Stratford Marshes. However, she continues that it remained mentioned as existing in the Abbey Lands Rate Book for 1837. Gardner also refers to a fire in 1847, of which no details had been found.[5]

Kemball Bishop

Illustration by Ann Vincent

## The Fire of 1802

In his research on Three Mills, Dr Keith Fairclough found evidence of a disastrous fire at Three Mills in April 1802[1]. The following report was published in the St James Chronicle, also known as, British Evening Post Thursday 8 April-Saturday 10 April 1802 and in the Morning Post on 9 April 1802[2]:

"The Flour Mills of Messrs. Metcalfe and Co at Bromley, near Bow, in Middlesex, were yesterday burned to the ground. The fire is supposed to have been occasioned by accident, but how, we believe, has not been ascertained. So early as three o'clock in the morning, the flames burst with such violence through the vent holes, for admitting the air to the grain, as to be visible at a considerable distance.

Every means that the place afforded were promptly exerted in attempting to extinguish the fire: Mr Metcalfe's engine was brought to bear upon it, and the people of the neighbourhood crowded to the spot. The fire had, however, got too much head to be subdued by any possible exertion. Part after part fell in, until nothing at length remained of these complete and excellent premises, except the bare shell. Of a very large stock of grain and flour, which they contained, scarcely any has been saved; it was either totally consumed, or so damaged as to be unfit for any use. The fury of the flames, and the force of the floors as they fell in, carried some of it into the river Lea: what remained within the walls presented the appearance of a black smoking dunghill, mixed with ashes and cinders of the timber. Without taking into account the great quantity of stock that has been destroyed, the loss is very considerable as the premises were new, and in the most perfect state. The greatest apprehensions were entertained for the safety of the adjoining houses. Fortunately the strength and thickness of the walls prevented the fire from extending to them and they escaped without damage. No lives were lost, not have we heard of any personal injury. While we sincerely regret so serious an evil, we feel pleasure in stating that this is the whole extent of it.

A report generally prevailed in the course of the morning that three Flour Mills, near Bow, had been set on fire by a mob the preceding night. This being represented to the Secretary of State, in a manner that could hardly leave a doubt of the truth of it, the military were ordered out, but on Sir Richard Ford, and some of the magistrates of the Police-Office proceeding to the spot, it appeared that the Mills had been burnt down entirely by accident, and that not the smallest disturbance whatever had taken place, as the magistrates were returning, they met a large party of the Horse and Foot-Guards going towards Bow, whom they dismissed with an assurance that their aid was not wanted, everything being in a state of greatest tranquillity."

Plaque dated 1776 still exists on the south face of the House Mill
Photo © Paul Grant

below:
18thC fire engines
Illustration by Ann Vincent

The Times of 9 April 1802 also reported briefly:

*"Yesterday morning between three and four o'clock, a fire broke out in a mill belonging to MR METCALF, at Bromley Essex. A great many fire engines were sent off from town, but before they arrived the mill was consumed. The adjoining premises, which consisted of two other mills, and valuable dwelling houses, were saved."*

This report raises a number of interesting points. First, the reference to *"Mr Metcalfe's engine"* suggests that the Mills/Distillery had their own fire-fighting equipment. *"The great many fire engines sent off from town"*, referred to in The Times may well have belonged to Metcalfe's insurance company. Unfortunately, the insurance company claims records for the period no longer survive.

Secondly, if the Mill was built in 1776, it is a little surprising that the premises should be described as *"new, and in the most perfect state"*. Dr Fairclough notes that Metcalfe took out an insurance policy with the Hand in Hand company in September 1795 for £2000 brick and £2000 timber for *"a corn mill 4 stories high on north side of road known by Three Mills"*, which clearly refers to the House Mill. In his note, Dr Fairclough wondered whether this implied that the House Mill had suffered a previous fire. If so, as with the 1802 fire, the south-facing brick wall, with the plaque dated 1776, must have survived.

Thirdly, the report does not name the mill which was subject to the disastrous fire of 1802. It has been suggested that *"the fire consumed the building on the site of the Clock Mill"* [3] (the latter was not built until 1817). However, several references in the report suggest it was the House Mill which burned down. First, it says the walls survived, as does the 1776 wall of the House Mill. The walls of the mill previous to the Clock Mill were of wood. Secondly, it refers to *"adjoining houses"*. There is no record of houses adjoining the pre-1817 mill on the site of the Clock Mill, whereas the House Mill derives its name from being built between two houses. Evidence of carbonised brick on the inner faces of the surviving walls confirm that the House Mill was seriously damaged by fire.

In any event, any dispute was resolved by dendrochronology in 2000. As part of the work to restore the House Mill, core samples were taken from beams and posts on each floor. The tree rings were analysed by the University of Sheffield and showed that the timber was Scots Pine which came from northern Poland; and that the timber was cut on varying dates between 1794 and 1802, much of it being between 1800 and 1802.[4] This showed conclusively that it was the House Mill which was burned to a shell in 1802; and it must have been quickly rebuilt behind the original 1776 wall.

below:
Painting of Three Mills before the fire in 1800
Reproduced by kind permission of the Nicholson family.

## Gladys and May Chandler's Diary

A Diary kept by the daughters of the Distillery Manager, who lived in the house to the east of the House Mill, from 1886 to 1927, when they moved from Three Mills, contains a number of references to events at Three Mills.[6] The Diary refers to an accident in the Mill on 31 January 1901, which covered the garden with maize, which suggests there may have been an explosion. It also refers to two fires, one which burned down Emery's on 14 October 1901 and a fire at Webster's on 20 February 1903. These incidents were not reported in the Stratford Express and neither Emery's nor Webster's has been traced. The Diary also refers to two other incidents: a very high tide at 7pm on 30 December 1904, which flooded the drawing room; and a new bridge over the river which fell down on the evening of 11 April 1905. Again, neither incident was reported in the Stratford Express.

William Chandler   Gladys & May

Photographs of the Chandler family supplied by Rosemary Elsmore

**Fire in 1908**

The Diary also refers to *"the dreadful fire [which] occurred at the Distillery [on 29 June 1908] when poor Mr Eddy was burnt to death, also a man named Green."* The Stratford Express reported that William James Chandler, spirit clerk at the Three Mills Distillery, gave evidence to the Coroner's Court on 1 July 1908. It said that at 9 pm, Mr Chandler:

*"went with the deceased [William Eddy, an Excise Officer] and a man named Green to the receiving room to take a sample of spirits. There were three vats in the room, two holding 5,000 gallons each, and the other 3,000 gallons. The trapdoor was replaced after the sample had been taken and it was the duty of the Inland Revenue officer to lock it. The vat where the explosion took place was about a third full. He [Mr Chandler] had just got outside the door, when he heard an explosion and saw a flash. He went back into the room, but by that time it was a mass of flames. He saw Green walk out from the flames. He helped him outside and he said 'Mr Eddy's inside'. Witness went in as far as he could, but by that time it was a mass of flames. ... Another explosion occurred while the witness was inside."*

Patrick Graham points out that the explosive range of ethyl alcohol lies between 3.3% and 19%; if the inspection was carried out with a naked flame, it could quite easily have caused an explosion, which would have been more dense than air and so flowed down towards the ground.

The photograph below, from the family collection of Rosemary Elsmore, who is the granddaughter of William James Chandler, is understood to be of the 1908 fire. The existence of the railway bridge in the background and the alignment of the buildings and the water suggest that it was on the spit of land between the mill tailraces and the Lee Navigation. E M Gardner refers to this as the Spirit Receiver Building.[7] This title and the involvement of the Excise Officer clearly refer to the storage of spirit after the distilling process and the building seems to have been located on the same site as the Bonded Warehouse bombed in 1940 (see below).

left:
Mr William Eddy, the excise officer who perished in the fire
© London Borough of Newham

Photographs of the 1908 fire supplied by Rosemary Elsmore

Gladys and May Chandler's diary also included experience of Zeppelin raids during the First World War. One flew close to Three Mill Lane on 8 October 1915 and a bomb landed on a bus. On 13 October 1915, May had to walk home from Bow Road station, as the railway had stopped because of the raids. On 3 September 1916 she saw one on fire from her kitchen window; it was brought down at Cuffley. On 14 September, two Zeppelins were brought down after dropping bombs on the Black Swan, Bow Road and Hogarth's, Bromley by Bow Bridge

Mr Eddy was a deacon at Carpenter's Road Baptist Chapel. Alfred Elias Green, aged 35, a labourer at the distillery, also died as a result of the burns he received while in the receiving room. Other witnesses at the two inquests included *'Charles Drake, the manager of the distillery'* and David Bird, who lived in Three Mill Lane and *'was in his back garden when he heard the explosion'*. In the 1901 Census, David Bird, aged 45, distiller's labourer, lived at 45 Three Mill Lane, with his wife Elizabeth and 6 children.

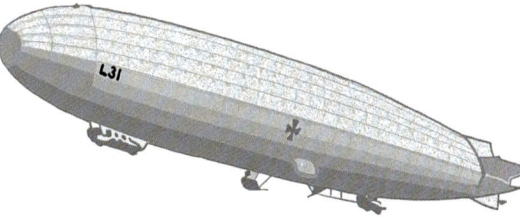

Illustration L31 Zeppelin
by Ann Vincent

The Diary also records a fire in the Tun Room on 26 September 1918, but damage was only about £20. No trace has been found in the Stratford Express.

## Granary Fire 1920

The Diary also records a Granary fire on 25 July 1920, which caused damage valued at £30,000.
*East End News* reported on 27 July:

"There was tremendous excitement during the early hours of Sunday morning in the Borough of Poplar. At about half past four o'clock it was found that the extensive distillery of Messrs. J. W. Nicholson and Co., of Bromley, was on fire. The outbreak, which was discovered by the night watchman, originated in the granary, a four-storey building, and the flames took a good hold before the fire brigade were able to bring them under control.

The West Ham and London County Council fire brigades were called, the building being actually in the West Ham area but close to the London boundary. The distillery comprised buildings of two and three floors, 240ft. by 170 ft. used as grain stores, malt mills and drying kilns. The fire had gained considerable hold, and before the brigade could get it under control half of the premises had been severely damaged. One of the walls of the building which overhangs the canal collapsed, bringing the roof with it, and a barge filled with grain was sunk by the weight of the debris. Between 5,000 and 6,000 quarters of grain was stored in the granary, and much of it was destroyed, but the fire was prevented from spreading to the whisky [sic] tanks.

The inhabitants of the neighbourhood were much alarmed, for the fire was at one time very threatening, and it was feared it would extend to adjoining premises. As it was, some damage was occasioned to neighbouring property, but the firemen succeeded in their efforts to confine it to the distillery premises."

Photograph taken after the fire, captioned "Three Mills Distillery Granary Fire, August [sic] 1920", ©Tower Hamlets Local History Library and Archives.

It is not clear whether the photograph is also from the East End News or was taken independently.

Recorded in the West Ham Fire Brigade minutes 1920-21 page 1388.

*"An extensive outbreak of fire occurred at Three Mills Distillery, Three Mills Lane, at 4.34 o'clock on the morning of July 25th. whereby a range of brick and timber buildings of four floors of about 210 by 240 feet, used as a granary, cooling rooms, malt mills, vats and kilns with contents was severely damaged by fire and water. A barge of 80 registered tons, lying in the River Lea at the side of the wharf, was sunk by falling brickwork. The estimated damage caused is £100,000."*

The damage seems to have been three times that in the press report. It does not comment about the attendance of the London FB but they would be there given the location.

Although the report refers to the building on fire overhanging the canal, it appears to have been sited to the east of the Mill tailraces. The building at the extreme left (just in the picture) appears to be the Clock Mill. The location of the barge and the boiler chimney at the rear appear to confirm this – the photograph was taken from the west or south-west. This can be compared with two photographs given to the River Lea Tidal Mill Trust by Jon and Catherine Battell (grand-daughter of the last brewer/distiller at Three Mills, Richard Ernest Blundell), which are clearly of the same building This is confirmed by the small building or room (possibly a lucam, or covered external hoist?) at right angles on the second floor. The drying kilns in the background of the second photograph confirm that the building was to the east of the mill tailrace, not the canal [ie the Lee Navigation]. This is also confirmed by a caption added by E M Gardner to an aerial photograph of 1938, which refers to *"The Circular Kilns Behind the Destroyed Granary"*.[8]

above right:
Destroyed granary
photos supplied by Catherine Battell

right:
Painting of Three Mills showing location of granary
© London Borough of Newham Archives

The photographs are undated but the style of the firemen's uniforms suggested that it was likely to be 1920 rather than the earlier fires recorded in May and Gladys Chandler's diaries.

## The Blitz

The Stratford Express reported on 18 October 1940:

*"Many fires were caused by incendiary bombs in a London Borough on Tuesday night, and the fire services were engaged for many hours in combating the flames while at intervals high explosive bombs burst around. Most of the fires occurred in houses, and these were quickly dealt with, but some industrial premises and shops were also set alight. Among the industrial premises that suffered were a distillery, warehouses and a bottling store. High explosive bombs also caused the destruction of a number of houses, as well as damaging a cinema, which has been closed for a long period. It is reported that casualties were slight."*

This is typical of reports in the period, which gave no details of bombing and under-played how serious they were; nor did it indicate casualties. Research is also complicated by the fact that the west side of the river was in the area of the London County Council, with records now at the London Metropolitan Archives, while Three Mills was on the Essex side of the river, with records at the Essex County Record Office at Chelmsford, each with separate Fire Brigade and Air Raid Precautions records; The records of the West Ham Fire Brigade, who would have been expected to be involved, do not refer to the Three Mills fire.

The bombing took place on 15 October 1940, which was a bad night for London's East End. The West Ham Civil Defence records report 28 high explosive bombs, 2 oil bombs, 43 incendiary bombs and two unexploded bombs falling between 5.30 pm on 15 October and 5.30 am on 16 October.[9] This includes factories and vital installations at North Woolwich, as well as the so-called Nicholson's Gin Works, Three Mill Lane, 3 shops and 46 houses. A separate report records 55 casualties, of which 12 were fatal, but does not distinguish Three Mills from other incidents.[10] The Poplar Bomb Register, which timed the occurrence at 20.28, says there were no casualties at Three Mills.[11] Reports from the LCC side of the Lea indicate that bombs fell there that night as well, when *"a number of incendiary bombs burned out on [the] road"* in Three Mill Lane.[12]

The bombing was also recorded in a number of secondary sources. Remembering Three Mills refers to the recollections of the late Cyril Demarne on the events of that night.[13] In private correspondence, he said:

*"My recollection is that, by and large, the distillery escaped major damage compared with that suffered by surrounding premises. I remember the incident … about a fireman's reaction to his night's activity. The fact that his appliance was ordered on to another fire in the dock area indicates that the distillery fire had been extinguished."*[14]

The latter statement appears to be inconsistent with the experience of Kemball Bishop (below). Another book reports that *"a warden crossing Bow Bridge notices that the river Lea was burning with a blue flame and realised that the local gin factory had been hit"*.[15] Unfortunately, the author relied on secondary sources and the original reference has not been traced.

Kemball, Bishop & Co Ltd, whose chemical works was on the west side of the Lee Navigation and south of Three Mill Lane, produced a record of their own experiences at Three Mills during the Second World War.[16] It includes the following graphic description of the events of 15 October 1940:

*"On the night of October 15th, an H.E. bomb was seen falling diagonally between our two chimneys. It fell on a warehouse of our neighbours across the river [ie the bonded warehouse of the distillery]. It blew half the roof towards our buildings. The masonry crashed on to an empty barge and sunk it. At the same time it blew down a dividing wall with its steel door in our neighbour's warehouse, exposing a huge stock of brandy and other spirits some contained in large vats supported on steel joists. The vats caught fire, and the vat seams could be seen opening, the brandy alight running down the sides. It was soon evident that the outer wall would crash into the river. The wall was supporting floor beams. As the wall fell the beams tilted and casks of brandy were seen sliding into the river. By this time the towing path and the surface of the river for 60 yards was ablaze with burning alcohol. The cabin of our coal crane, 40 feet in the air, caught alight by radiation. In two minutes of the dropping of this bomb and the starting of our neighbour's fire, our A.R.P. men were playing water across the river into the burning mass. We then had to concentrate on the protection of our own buildings. Our woodwork was catching fire, also by radiation. No man because of the heat could get on the wharf. Thrown out sandbags, their coverings soon in ashes, protected our men who worked throughout the night spraying our buildings with water. In the morning 12 barrels of brandy were rescued from the river. The river was blocked with sunken barges and masonry. It is stated that the market value of the spirits destroyed was in the neighbourhood of £1,000,000."* [17]

It is interesting that Kemball, Bishop refer throughout to 'brandy', which was not known to have been distilled at Three Mills at this time.

Illustrations by Ann Vincent

## The Blitz cont.

A visitor to the House Mill several years ago said his father loved to recall visiting the site the next morning, where, he said, he saw casks of alcohol still floating in the river and the Excise men in rowing boats smashing them up. However, this does not seem consistent with the end of the Kemball, Bishop report.

Audrey Stenhouse was the daughter of Richard Ernest Blundell, who was the brewer/distiller at Three Mills from 1932 until after the bombing. Mrs Stenhouse recalled the bombing of Three Mills on 15 October 1940, when she was 17. It was a moonlight night and incendiary bombs had fallen. They had covered them with sandbags, but one was a bag of coal which caused a fire! During the raid, the family took refuge in the large air raid shelter in one of the distillery buildings, where there were bunks and room for all the staff and their families. She recalled the screaming of the bombs. When they emerged, it was still possible to gain access to the house to retrieve belongings, though it was never lived in again. They moved for a short while into a dwelling between the east end of the mill and the river (the Custom House?). Mrs Stenhouse then moved away to Hither Green and never returned to Three Mills.

E M Gardner records that the *"dwelling house on the left"* [ie to the west of the House Mill] *"was destroyed and that on the right so badly damaged as to be unsafe"*. [18] The late Ernie Springall's drawing shows his recollection, many years later, of the destroyed house to the west of the Mill. The house to the east was demolished in 1954.

Kemball, Bishop also recorded another incident, some 6 weeks later, which did not directly affect the distillery:

*"... on November 29th, 1940, while an air raid was actually in progress ... an H.E. bomb had been seen falling within a few feet of our premises but no explosion followed. Some time later one of our people coming on night-shift reported that he had fallen over a projection in Three Mill Lane. It was the H.E. bomb protruding 1½ feet out of the ground and unexploded ... It was found that the bomb had fallen upside down, its dislocated fuse at the top of the column, and thus we again escaped by a miracle."* [19]

Mrs Stenhouse recalled seeing the unexploded bomb in the roadway of the bridge across the Navigation. This incident was recorded in the Poplar Bomb Register:

*"POP 965 28.11.40 altered to 29/11 Outside Nicolson Bondage Store [sic] Three Mill Lane. ......Casualties – Nil. Position of any unexploded bombs – Three Mill Lane Bridge. Time of Occurrence 20.00. Telephone wires down. Bomb is visable [sic]."*

The West Ham Bomb Register, as well as record the bombing on 15 October 1940, notes a hit at *"Nicholson's Factory (Distillery)"* on 16 February 1944 by an unexploded AA shell.

Drawing from memory of the destroyed Brewer's house
By the late Ernie Springall

## Fire 1947

The Evening Standard reported on 17 October 1947, Two men injured: 2000 gals. spirits burned":

"Two men were injured and 2000 gallons of spirits were destroyed in a fire at Nicholson's Distilleries, Three Mill-lane, Bromley-by-Bow, today.

"In St Andrew's Hospital, Bow are Jack Sodo, of Caistor Park-road, Plaistow, severely burned, and Richard Woodcock, with slight burns and shock.

### Drums explode

"Men were unloading tons of spirit from a barge at a wharf alongside the distillery when it is believed a spark caused a drum of spirit in the spirit stores to explode.

"Within a few moments other drums were exploding and flames were shooting 30ft. out of the windows of the three-storey disused distillery which was being used as a spirit store.

"Twelve NFS pumps went to the distillery and dense clouds of smoke covered the area.

### Like a torch

"Jack Sodo had his clothes set on fire, and he ran like a human torch into the distillery yard. On his way to hospital he kept saying he hoped money he had collected in a football pool among his fellow workers was safe.

"The alarm was raised by Charles Tyne, 60-year-old labourer. As soon as the first drum of spirit exploded he shouted to nine other men working in and around the store, "The place is alight; get out quick!".

"Two hours after the fire started the firemen were still playing their hoses into the building, and spraying water on the outside walls, to cool them."

## Unexploded Bomb 2008

In June 2008 an unexploded bomb was found in the Prescott Channel, which is the eastern boundary of Three Mills Island, during works to build the new Three Mills Lock . The Winter 2008 Newsletter of the Friends of the House Mill recorded:

"The whole area, including the House Mill, Miller's House and the Distillery/Film Studios, had to be immediately evacuated. The Bomb Disposal Squad, the Royal Engineers, worked around the clock over the next five days to deal with the threat, successfully we are glad to say. It is understood to have been a 2500lb bomb which was not detected when the river was scanned prior to the works; presumably it had been buried behind the concrete reinforcements. Had it exploded, either when it was dropped or when it was being disarmed, we understand it could have destroyed not only 4,000 houses, but probably also the [Distillery buildings and the] House Mill.
Indeed, when the controlled explosion took place, there proved to be more explosive remaining than had been expected – there was slight damage to the roof of the Distillery/ Film Studio buildings and to the roofs of several parked cars."

Unexploded bomb
Photo © Carl Ainley

## *Acknowledgements*

The Trust are grateful for the assistance of staff at the London Metropolitan Archives, the London Borough of Newham Archives and Local Studies Library and the London Borough of Tower Hamlets Local History Library and Archives. I am also grateful to Jon and Catherine Battell and Rosemary Elsmore for the donation of illustrations; and to Ken Drew for drawing my attention to the material on the 1920 and 1947 fires.

[1] Fairclough, Dr K R Unpublished note (undated).
[2] Reports of the Fire also appeared in the *Lloyds Evening Post and British Chronicle*, Wednesday 7 April-Friday 9 April 1802; and the Times Friday 9 April 1802.
[3] Hitchmough, Wendy, *The House Mill – An Archaeological Record – Section 1 Historical Documentation* Julian Harrap Architects 1998, page 1.35.
[4] University of Sheffield Dendrochronology at The House Mill Letter dated 5 July 2000.
[5] Gardner, E M *The Three Mills Bromley-by-Bow* originally published by the Society for the Protection of Ancient Buildings 1957, reprinted by the River Lea Tidal Mill Trust, p. 25.
[6] *Remembering Three Mills*, River Lea Tidal Mill Trust 2008, Appendix 2.
[7] Gardner, E M op cit page 26.
[8] Ibid, Plate 3 – 1938, page 21.
[9] Essex County Record Office, *West Ham Civil Defence Records* C/W 2/1/49.
[10] Essex County Record Office, *Essex CC Group 7 (LCDR) Metropolitan Essex Operational Report from 1939 to June 1944* C/W 2/3/1, 2.
[11] Tower Hamlets Local History Library *Metropolitan Borough of Poplar – Bomb Register* POP 946 Ref 2 15.10.40 20.28 Nicholson Gin Factory [sic], Three Mill Lane
[12] London Metropolitan Archives, *London Fire Region Fires – Twice Daily Situation Reports (LCD Region)* FB/WAR/LFR/2/26.
[13] *Remembering Three Mills*, page 10, quoting from Demarne, Cyril *The London Blitz – a Fire man's Tale* Parents Centre Publications 1980, page 34.
[14] Private communication 17 September 1996.
[15] Palmer, Alan *The East End: Four Centuries of London Life* page 143.
[16] *A Miracle at the Crown Chemical Works*, Kemball Bishop & Co Ltd, Tower Hamlets Libraries Local Collection LC 461, Class 6706.
[17] Ibid, pages 11-12.
[18] Gardner, E M op cit, page 26.
[19] Kemball Bishop & Co Ltd, page 12.

Illustration by Ann Vincent